STUDIES IN CONTEMPORARY EUROPE

In the past quarter of a century European society, and Europe's relations with the rest of the world, have been radically transformed.

Some of these changes came in the wake of the Second World War; others – and in particular the division of Europe – followed as a result of the Cold War. In addition, throughout the period other forces, and especially technological change, have been at work to produce a major recasting of the fabric of European society and Europe's role in the world. Many of these changes, together with their attendant problems, have transcended the political and economic divisions of the continent.

The purpose of this series is to examine some of the major economic, social and political developments of the past twenty-five years in Europe as a whole – both East and West – considering the problems and opportunities facing Europe and its citizens today.

A farm is a farm is a farm!

Seen one. See them all

what's the point in a whole book about a farm

25.6.72

STUDIES IN CONTEMPORARY EUROPE

General Editors: ROY PRYCE *and* CHRISTOPHER THORNE

Published titles

AGRICULTURE	HUGH D. CLOUT
RURAL SOCIETIES	S. H. FRANKLIN
YOUTH AND SOCIETY	F. G. FRIEDMANN
EDUCATION	JOHN VAIZEY

In preparation

EUROPE AND THE THIRD WORLD

SOCIAL STRATIFICATION

THE STRUCTURE OF INDUSTRIES

THE URBAN EXPLOSION

POPULATION MOVEMENTS

ECONOMIC PLANNING

THE MASS MEDIA

CHRISTIAN DEMOCRACY

WOMEN IN SOCIETY

SOCIAL DEMOCRACY

PATTERNS OF CO-OPERATION AND INTEGRATION

THE QUEST FOR ECONOMIC GROWTH

AGRICULTURE

HUGH D. CLOUT

Department of Geography, University College London

MACMILLAN

First published 1971 by
THE MACMILLAN PRESS LTD
London and Basingstoke
Associated companies in New York Toronto
Dublin Melbourne Johannesburg and Madras

SBN 333 12293 3

Printed in Great Britain by
THE ANCHOR PRESS LTD
Tiptree, Essex

CONTENTS

LIST OF DIAGRAMS

LIST OF TABLES

PREFACE

In 1945 European agriculture was faced with the major problem of increasing food production to satisfy her population after years of war-time deprivation. Twenty-five years later the continuing application of science and technology has placed European farmers in the novel predicament of having to cut back production and dispose of surpluses. Governments have intervened in agricultural organisation with differing degrees of ruthlessness in both Communist and non-Communist countries. They have met with differing degrees of success when the results are considered from political, economic and social points of view. This essay attempts to explore agricultural change in post-war Europe through the particular medium of structural planning and finally points to the problem of trying to decide what should replace farming as an employer of labour and a user of land in the 1970s and beyond.

ACKNOWLEDGEMENT

I am indebted to Margaret Thomas for her skilful preparation of the illustrations.

Department of Geography,
University College London,
September 1970

HUGH D. CLOUT

1. AGRICULTURE IN THE CHANGING EUROPEAN ENVIRONMENT

Economic and social life in Europe since 1945 has been characterised by rapid urbanisation and an expansion of employment in manufacturing and tertiary activities.[1] Land which was formerly used for farming has been claimed for factories, roads and housing estates. People who worked the land or might have been expected to do so have moved to better-paid jobs in towns and cities. Advances in science and technology have been appled to all sectors of economic and social life, but, with certain notable exceptions, farming has been subject to more constraints and has been slower to respond than other activities.

In many respects agriculture looks like a loser, but in fact post-war farming has been characterised by increasing yields and the continuing substitution of capital for land and labour resources. Scientific advances in agriculture which might theoretically be possible have been subject to many complex constraints stemming from the political, historical, social and economic environment of farming. Governments throughout Europe have intervened to try to overcome these constraints and remodel farming to attain a number of objectives which are sometimes of an ambiguous or even a contradictory nature. It is with these forms of intervention that this essay will be primarily concerned.

THE DIVERSITY OF EUROPEAN AGRICULTURE

So far only broad generalisations about farming have been made, but unfortunately this is a grossly unrealistic approach since there are many more agricultures in Europe than there are nations, each with its own characteristics at any given moment in time and each sharing in and contributing to the changing pattern of European farming over the past quarter of a century. Agricultural diversity results from numerous factors, of which only a few may be mentioned here. First, there is the complex physical environ-

[1] See the essay in this series by T. B. Bottomore on *Social Stratification*.

ment of Europe, with soils and climatic conditions ranging from the arid Mediterranean to the Arctic Circle and from the cool, humid Atlantic coastline to the vast continental interior of the U.S.S.R. A general response to this diversity is suggested by the latitudinal crop belts shown in Fig. 1.

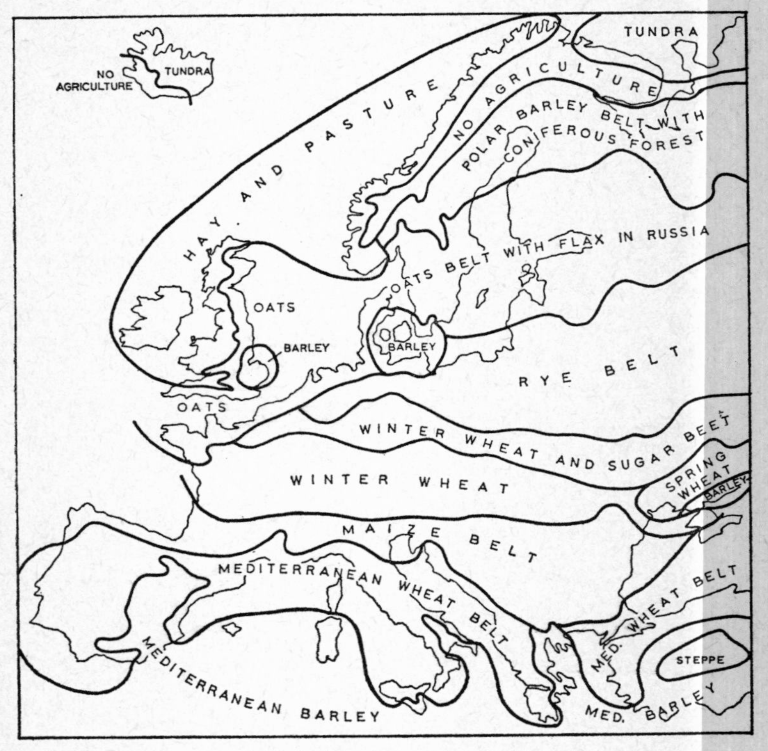

FIG. 1 European crop belts

Several other important factors must be superimposed on this simple picture to give it more reality. Micro-features include the varying suitability of specific sites for producing particular crops, and differences in the technical and intellectual ability of farmers to practise modern agriculture. Major differences between countries resulted from the stage of general economic development reached by each region and the nature of the political, economic and social systems which operated there. The net result was that immediately before the war Belgium, the Netherlands and Denmark, with small and medium-sized family farms, had the highest levels of output per man in farming, together with the United Kingdom and the eastern part of Germany with larger properties and greater social stratification. The lowest levels of income per farm worker were recorded in Spain, Portugal, Hungary and Italy, with large landed estates and many landless farm labourers, and in the Balkan countries where small peasant farms predominated. These basic inequalities were to be evident throughout the post-war period.

Further differences immediately after the Second World War resulted from the degree of destruction and devastation which had been experienced during the war years. An additional factor which was soon to create further diversity was the different nature of government intervention in agricultural management, ranging from land reform and collectivisation as part of major social reorganisation by the new Communist regimes of Eastern Europe to less thoroughgoing forms of intervention in the West.

POST-WAR RECOVERY

Immediately after war ended, the main concern of European farming was to regain production levels and to feed the people of Europe. In several Western European countries production was only two-thirds of what it had been six years earlier, and in West Germany, Austria and Greece it was well below the two-thirds mark. Even by 1947–8 farm production in Western Europe was no higher than it had been in 1910. Agricultural recovery in the West, however, was rapid and pre-war levels were reached in 1948–9 for Sweden and the United Kingdom and in 1949–50 for all other Western nations save West Germany, Austria and Greece. After this initial phase of remaking lost ground there was a further rapid increase in output in the second phase of post-war agricultural development which pushed production in Western Europe well beyond pre-war levels and did not begin to slacken until 1954–5.

This rapid recovery was produced by resuming normal farm practices and adopting improved methods rather than by restoring land to agricultural use. Only in the most battle-torn areas had land actually fallen out of cultivation, but the quality of management had declined severely during the war. In many places both during and immediately after the Second World War humans competed directly with animals for the consumption of limited supplies of cereals. Given these shortages of livestock feed, meat and milk yields were low. However, all problems except the depletion of livestock numbers were overcome rapidly in Western Europe as men returned to the farms and investment in agricultural development restarted.

By contrast, recovery took very much longer in Eastern Europe where there had been severe losses of livestock and much land had been very poorly farmed as a direct result of war action and the enforced movement of the resident population. In Yugoslavia, for example, 80 per cent of farm equipment, 60 per cent of draught animals and 40 per cent of rural housing had been destroyed during the war. Drought throughout Eastern Europe in 1946–7 worsened conditions and led to a further slaughter of animals because of insufficient foodstuffs. The war-time shortage of manpower and fertilisers incurred long-lasting damage to the soil. Later upheavals associated with land reform and collectivisation meant that as late as 1950 agricultural production in Eastern Europe remained below the 1935 level. Numbers of cattle, pigs, sheep and horses in Czechoslovakia, Hungary, Poland and Yugoslavia totalled only 56·2 million in 1950, which was well below the 1935 figure of 67·2 million. Milk production in the same four countries had fallen from 19·6 million metric tons to 14·4 million metric tons and production of wheat, maize, rye, barley and oats had toppled from 32·9 million metric tons to only 28·0 million metric tons.

By 1950 differences in agricultural development which had been marked in the 1930s were still apparent (Tables 1 and 2). The United Kingdom was at a very advanced stage (and of course had not suffered war-time invasion), containing only 2 per cent of the European agricultural labour force but producing 9 per cent of its total output.[2] The value of output from each British farm worker was to reach $2310 per annum a few years later. Continental Europe could be divided into two parts according to stage of economic development and post-war recovery. The first comprised France, northern and central Europe, which

[2] Agricultural changes in the United Kingdom and the U.S.S.R. are not considered in detail in this essay, but some figures are included for comparative purposes.

Table 1

Agriculture in Europe (excluding the U.S.S.R.) in 1950

(percentages)

	Total area	*Arable area*	*Agricultural production*	*Population in farming*	*Total population*
United Kingdom	4	4	9	2	12
Northern Europe[a]	24	8	8	4	5
France	10	13	17	8	10
Central Europe[b]	8	8	19	11	20
Eastern Europe[c]	18	28	21	28	22
Southern Europe[d]	36	39	26	47	31

[a] Denmark, Finland, Iceland, Ireland, Norway, Sweden.
[b] Austria, Belgium, Federal Germany, Luxembourg, Netherlands, Switzerland.
[c] Albania, Bulgaria, Czechoslovakia, East Germany, Hungary, Poland, Romania.
[d] Greece, Italy, Portugal, Spain, Turkey, Yugoslavia.

Source: United Nations, *Economic Survey of Europe since the War* (Geneva, 1953), p. 164.

contained 23 per cent of European farm workers and produced 44 per cent of total output. The value of output from farm workers in these countries ranged from $2260 per annum in Belgium to

Table 2

Value of Output per Active Male in Farming, 1955

	U.S. $	*Index*		*U.S. $*	*Index*
United Kingdom	2310	100	France	1510	65
Belgium/Luxembourg	2260	98	Finland	1310	57
Denmark	2000	87	Austria	1140	49
Sweden	1905	82	Ireland	1110	48
Netherlands	1815	79	Italy	820	35
Switzerland	1745	76	Greece	570	25
Federal Germany	1685	73	Spain	435	19
Norway	1515	66	Portugal	385	17

Source: P. L. Yates, *Food, Land and Manpower in Western Europe* (London, 1960), p. 148.

$1110 in Ireland. The second zone involved Southern and Eastern Europe, containing 75 per cent of the agricultural work force but contributing only 47 per cent of total production. The value of output per worker (unfortunately available only for non-

Communist Europe) ranged from $820 in Italy to $385 in Portugal.

PROGRESS BEYOND RECOVERY

A decade later two significant changes in European farming could be discerned: the large-scale intervention of governments to promote agricultural development and the revolution in methods of cultivation and the equipment and structure of farm holdings. In Western Europe these changes had been translated through price-support measures and aid for technical development, but in Eastern Europe they had been associated with collectivisation.

Since 1945 unparalleled progress had been made in the technical and biological aspects of agricultural production. Thus, for example, highly productive or resistant crop strains had been developed, new breeds of livestock perfected, new fertilisers, agricultural chemicals and machines manufactured, all combining to reduce the quantity of land and labour resources required to produce a given quantity of foodstuffs. Progress had been so great in fact that agricultural objectives had to be reversed. The initial problems of making up lost ground and producing enough food to satisfy Europe's growing population gave way, in Western Europe at least, to trying to cut back some aspects of production and dispose of surpluses in the third phase of post-war farming development.

Statistics for wheat production are readily available and are less ambiguous than those for livestock and many crops. They therefore will serve as a basis for discussion and, although not fully representative of farm production as a whole, may be used in a very general way to suggest spatial differences in agricultural advance between 1948–50 and 1960–2. Taking Europe as a whole, but excluding the U.S.S.R., average wheat yields rose by one-half from 1410 kilograms/hectare to 2060 kg/ha and remained clearly the highest for any of the world's major regions (Fig. 2). Comparable changes in the U.S.S.R. involved a 20 per cent increase from 810 kg/ha to 970 kg/ha. Vast increases in total Soviet wheat production were achieved by doubling the wheat area during the Virgin and Idle Lands plough-up campaign (Fig. 2).

Rates of increase in wheat yields varied enormously between countries (Fig. 3). Decreases were recorded only in Albania from

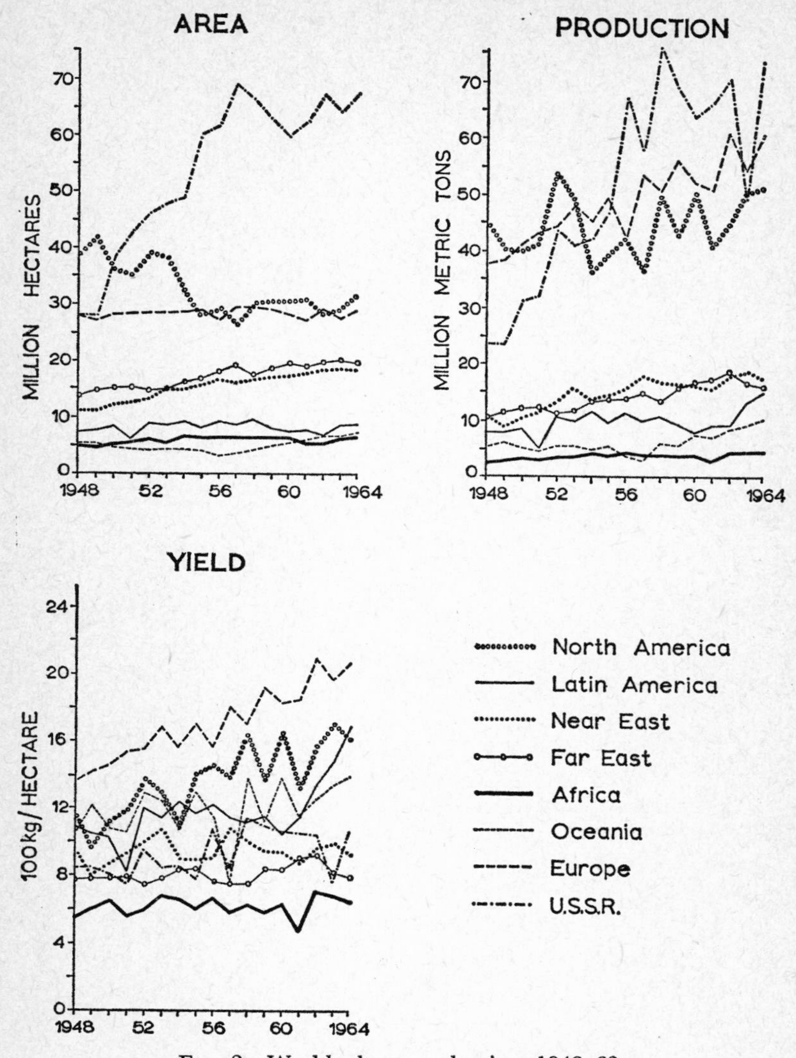

FIG. 2 World wheat production, 1948–62

960 kg/ha to 880 kg/ha. Seven countries experienced significantly above-average increases of more than 50 per cent. It is essential, however, to stress that large yield increases produced very different results, since wheat yields already varied enor-

mously throughout Europe in 1950. Increases of more than 50 per cent in Greece and Romania raised production to 1570 kg/ha and 1310 kg/ha respectively (Fig. 4), but, in spite of such dramatic increases, their yields were well below the European average

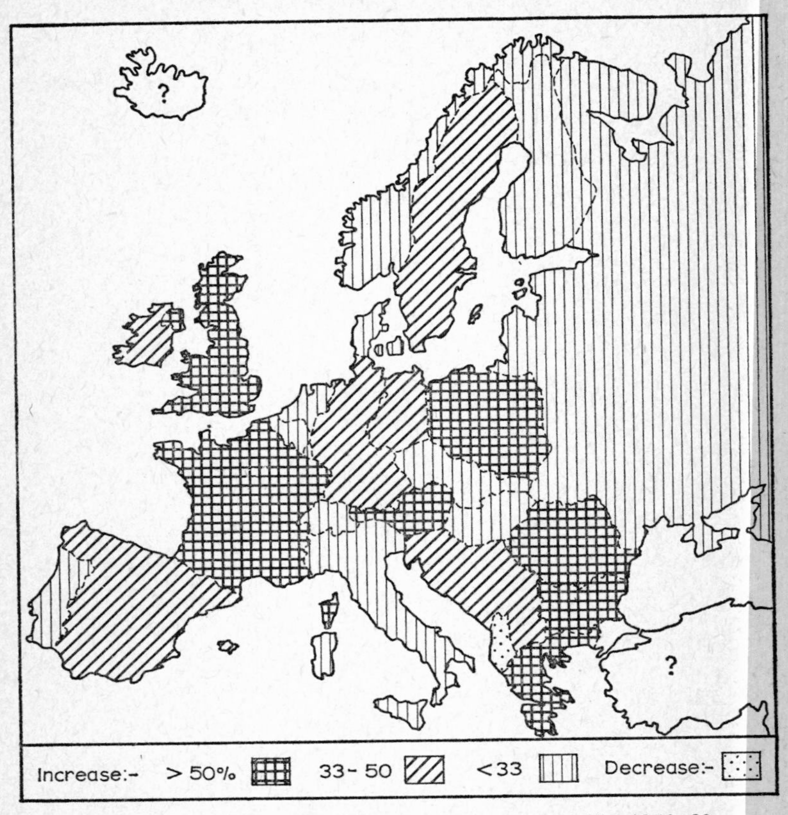

FIG. 3 Variation in national average wheat yields, 1950–60

at both dates. In the United Kingdom a 54 per cent increase raised wheat yields from 2680 kg/ha to 4130 kg/ha, but in the already highly efficient and intensive agricultural countries of north-western Europe increases of less than one-third produced exceptionally high yields of 4480 kg/ha in the Netherlands, 4030 kg/ha in Denmark and 3960 kg/ha in Belgium. These variations suggest that regional contrasts in agricultural development, already noted at earlier dates, continued to underlie the progress made after 1950.

Advances in wheat output and many other aspects of production were accomplished with a decreasing labour input, for which

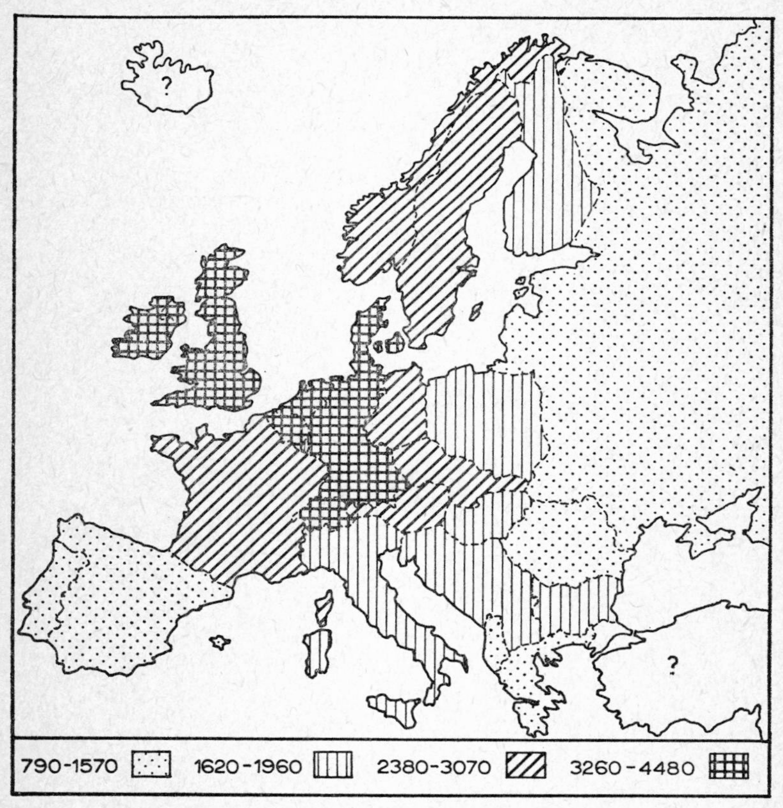

Fig. 4 National average wheat yields, 1960 (kg/ha)

machinery was being substituted. Thus between 1955 and 1965 the number of tractors in Europe (excluding the U.S.S.R.) increased from 1·77 million to 4·58 million and combine-harvesters from 96,500 to 557,000. The number of tractors in the U.S.S.R. doubled from 818,000 to 1·6 million and combines increased by one-half from 332,000 to 520,000. The agricultural labour force decreased everywhere in Europe during the 1950s save in Turkey

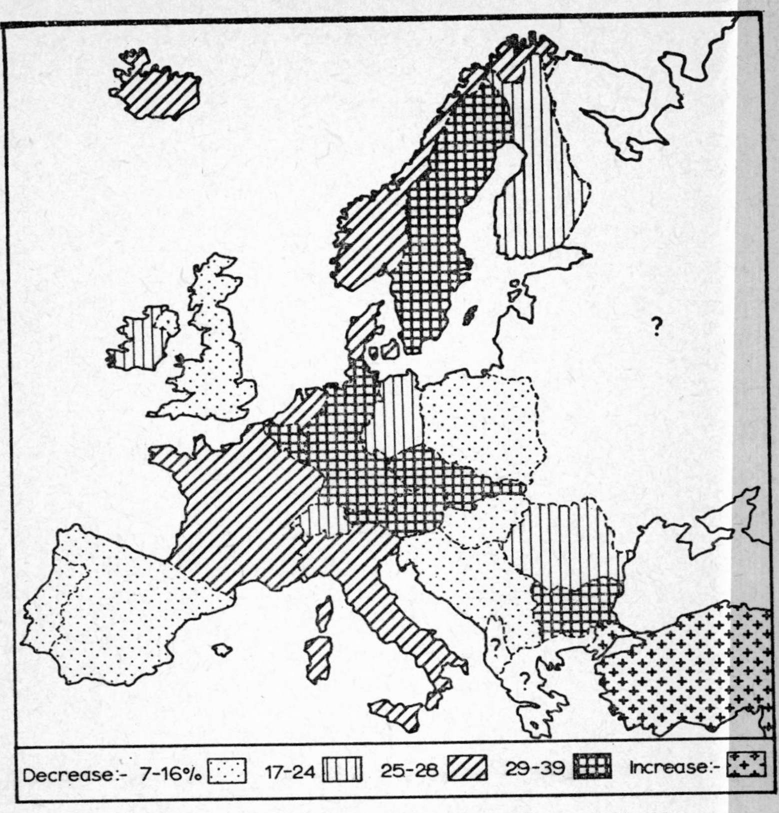

FIG. 5 Change in agricultural labour force, 1950–60

(Fig. 5). Elsewhere reductions of between 7 per cent in Poland and 39 per cent in Belgium were recorded. The smallest losses were: first, in Spain and Portugal whose national economies were still at a phase of development which north-west Europe had experienced prior to 1930 and where industrial alternatives to agricultural work were rare; second, in Poland and Yugoslavia where, as will be discussed later, the rejection of collectivisation allowed family farming to survive on small plots and kept large numbers of workers on the land; and, third, in the United Kingdom where the proportion of workers in farming was the smallest in the world. Half the countries of Europe witnessed reductions of 25 per cent or more in their farm labour force, with peak losses in Federal Germany (—35 per cent) and Belgium (—39 per cent).

In 1965 the less developed nations of Eastern Europe still had between two-fifths (Poland) and three-quarters (Turkey) of their labour force in farming (Fig. 6). In north-western Europe less than 14 per cent of national work forces were in agriculture. This was a recent stage in a long-established trend, since by the 1920s the peak numbers in agricultural jobs had already been recorded in many parts of north-west Europe. In 1950 farming employed 27 per cent of the labour force of non-Communist Europe. By 1955 it had fallen to 24 per cent and has since reached 18 per cent in 1970. Such a reduction in labour was desirable

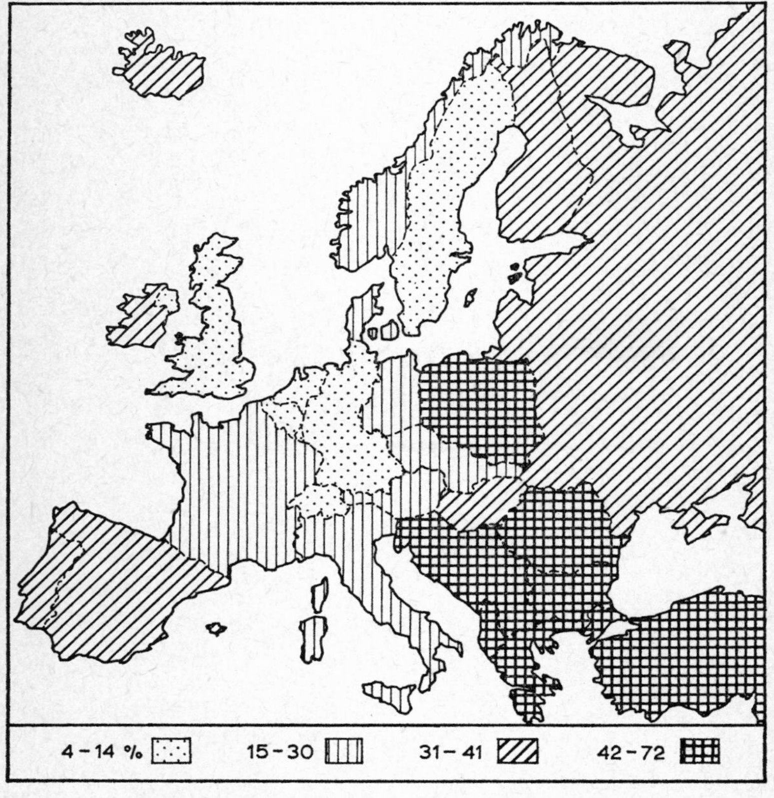

FIG. 6 Proportion of labour force in farming, 1965

and indeed essential if scientific progress was to be further applied to European agriculture. But changes in absolute numbers conceal increasingly unfavourable features in the quality of the agricultural labour force which, in its age and sex characteristics, does not represent a typical cross-section of the total population. In the Common Market countries 50 per cent of those employed in farming are over fifty-seven years of age, and in Eastern Europe the proportion of workers aged fifty or over is more than twice as great in farming as in other jobs (Table 3).

The explanation for this lies partly in the fact that farming in Continental Europe is predominantly a low-income activity when

TABLE 3

Percentage of Workers aged fifty years or more (a) in Non-agricultural Employment (b) in Agricultural Employment in Eastern Europe

		Both sexes		*Males*		*Females*	
		(a)	*(b)*	*(a)*	*(b)*	*(a)*	*(b)*
Bulgaria	1965	13	31	17	40	6	25
Czechoslovakia	1961	21	49	23	55	18	44
Hungary	1960	19	42	21	44	17	37
Poland	1960	18	38	19	40	16	36
Romania	1966	16	31	18	35	11	28
U.S.S.R.	1959	11	28	13	26	9	29

Source: United Nations, *Economic Survey of Europe in 1968* (New York, 1969), p. 213.

compared with industry or office work. It is certainly true that agricultural yields have risen remarkably since 1945, but rates of agricultural growth are below those for other sectors of the economy. Table 4 demonstrates this by comparing changes in Gross Domestic Product and Gross Agricultural Product for twelve Western countries between 1950 and 1962. Incomes derived from farming lag behind those from other sectors of the economy. The explanation for this is complex and because of feedback features it is not always easy to separate cause from effect. At one level, low incomes derive from the tendency for agricultural production to increase more rapidly than the aggregate demand for food. At another level they result from inadequacies in farm structures, from the continuation of farming in impoverished regions (with poor soils, harsh climate or inadequate communications) and from the frequent inability of farmers to take advantage of modern advances through personal or financial inadequacies or because of constraints imposed by the social and economic environment in which they operate. Anyway a vicious downward spiral is created in which the young and better educated with most drive and ambition leave the countryside for more challenging and rewarding urban jobs.

Another unsatisfactory feature, particularly pronounced in industrialising Eastern Europe where growing numbers of male workers are required for factory work, is that farming is becoming increasingly a female occupation, with numbers of women to every 100 men in agriculture ranging from 123 in Poland to 134 in Romania. The proportion of the total agricultural work force composed of women of all ages and men over fifty ranges from 65 per cent in Hungary to 70–75 per cent in all other East European countries. This is a highly unfavourable situation since the agricultural labour force in most European countries is excessively inflated by workers of low productivity who view farming as a way of life rather than a competitive economic activity and for whom subsistence is more important than efficiency. Columns (ii) and (iii) in Table 5 shows that in 1960 only the agricultural population of the United Kingdom and the Benelux countries

TABLE 4

Variations in Gross Domestic Product and Gross Agricultural Product at Factor Cost, 1950–62, 1954 prices

(1954 = 100)

	G.D.P.	*G.A.P.*
Austria	184·2	132·2
Belgium	119·0	116·6
Denmark	158·8	124·1
France	173·0	135·0
Federal Germany	233·5	131·0
Greece	197·6	174·7
Italy	205·2	138·5
Netherlands	169·7	130·5
Norway	154·5	95·0
Portugal	165·2	116·2
Turkey	169·8	143·9
United Kingdom	132·6	130·5

Source: Organisation for Economic Co-operation and Development, *Agriculture and Economic Growth* (Paris, 1965), pp. 102–3.

TABLE 5

Percentage of Working Population in Agriculture (i) in 1965 and (ii) in 1960; and (iii) *Percentage of Gross Domestic Product (Western Europe) or Net Material Product (Eastern Europe) derived from Farming in 1960*

	(i)	(ii)	(iii)		(i)	(ii)	(iii)
Albania	58	75	?	Italy	25	29	17
Austria	20	23	11	Netherlands	8	11	10
Belgium/Luxembourg	6	7	7	Norway	16	16	11
Bulgaria	44	64	24	Poland	42	47	24
Czechoslovakia	16	23	15	Portugal	40	41	26
Denmark	15	17	14	Romania	58	59	33
Finland	25	25	21	Spain	34	40	27
France	18	20	10	Sweden	12	16	8
East Germany	18	18	12	Switzerland	10	11	?
Federal Germany	11	13	6	Turkey	72	79	42
Greece	53	54	28	United Kingdom	4	4	4
Hungary	31	31	23	U.S.S.R.	33	34	21
Iceland	35	38	?	Yugoslavia	53	57	26
Ireland	32	35	25				

Sources: *World Atlas of Agriculture: Europe, U.S.S.R., Asia Minor* (Novara, 1969), p. 16; Food and Agricultural Organisation, *Production Yearbook* (Rome, 1968).

contributed to the Gross National Product at a rate which directly represented its own proportion of the population. Elsewhere in Europe the farm labour force was grossly inefficient, even though farming in Southern and Eastern Europe frequently contributed more than 25 per cent of the G.N.P. or Net Material Product (Fig. 7).

The move from farming to alternative employment is often not a sharp break but rather a gradual process in which two jobs combine.[3] Worker-peasants have full-time industrial or tertiary jobs but still live on their farms which are managed at weekends, in the evenings (five o'clock farmers) or by the farmers' wives and families. Estimates suggest that the proportion of holdings operated by worker-peasants ranges from 15 per cent in France to 30 per cent in Federal Germany and even higher in Switzerland, Poland and Yugoslavia. The worker-peasant situation offers a cushioning effect in times of industrial slump, provides higher incomes than farming would alone supply and helps to maintain services in rural areas. But worker-peasant farms have many defects. Normally they are very small (often less than 1 ha) and

[3] See the essay in this series by S. H. Franklin on *Rural Societies.*

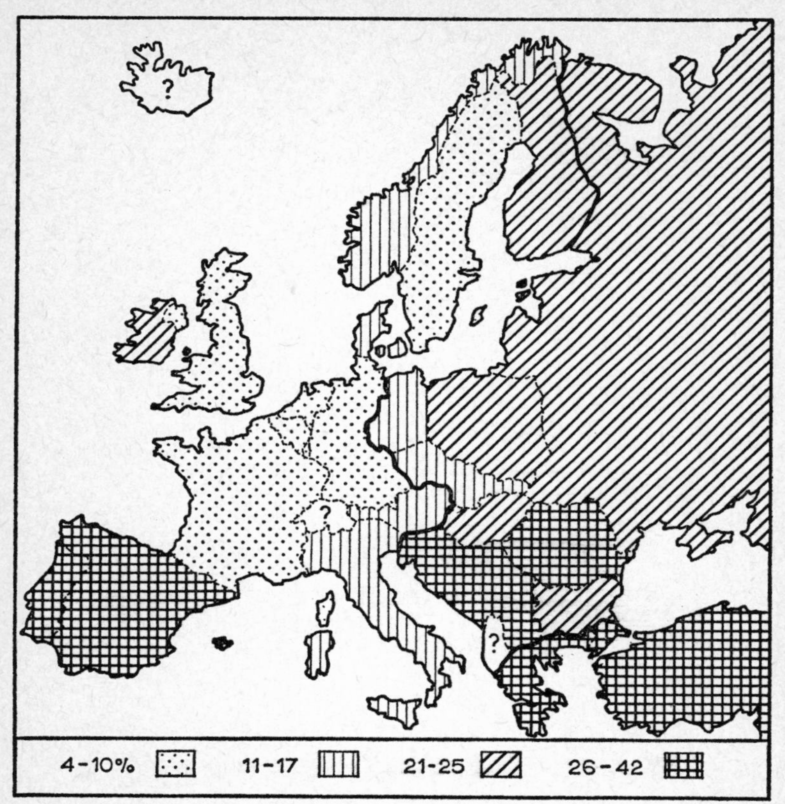

FIG. 7 Proportion of Gross Domestic Product (Western Europe) or Net Material Product (Eastern Europe) derived from farming, 1960

excessively fragmented. They are frequently over-mechanised and have very low productivity. Perfectly fertile and accessible land may fall out of cultivation simply because worker-peasant families lack the time or inclination to farm it.

FUNDAMENTAL AGRICULTURAL PROBLEMS IN MODERN EUROPE

From the foregoing paragraphs it is clear that 'agriculture' is an ambiguous term, referring at the same time to an economic activity which requires rationalisation and also to a way of life which not only supports efficient producers but shelters a host of inefficient, elderly and part-time 'farmers'. Europe's current agricultural problems are rooted in this ambiguity.

Impressive though figures on increased yields, mechanisation, reductions in the labour force and other indices of agricultural change undoubtedly are, there remains a marked gap between what might be possible if continually evolving production techniques could be applied to the full in European farming and what has in fact been achieved. The European farm labour force, although decreasing rapidly, remains the legacy of pre-mechanised conditions and is much too numerous and ill-trained in most countries for the needs of modern agriculture. Unfortunately the task of attempting to define its optimum size has received little attention.

Secondly, there is the constraint on agricultural progress imposed by the highly complex relationship between the farming population and the land it cultivates, translated through land tenure, farm size, field and farm layout and patterns of rural settlement. All these features undergo slow natural change but they still reflect the requirements of agriculture in the past. Many fields are too small for modern equipment to be used and often farms are below the minimum size necessary for viable modern production. The definition of minimum viable sizes is a highly complex issue, but the figures in Table 6 derived from independent national criteria suggest the large proportion of non-viable units. Farm labourers are more mobile than farmers and hence changes in the number and size of farm holdings are far slower than reductions in the total agricultural labour force (Table 7). Villages and hamlets frequently lack the schools, shops and other services which the agricultural population, exposed to mass media of communication, has learned to expect.[4] The whole structure of the European countryside is becoming more inefficient as each year goes by and scientific and technical advances are applied to farming. The scale of agricultural activity demands dramatic enlargement.

Thirdly, there is the fundamental problem caused by the continuing influence of traditional methods of crop and livestock production which no longer correspond to current consumer de-

[4] Case studies of social change in rural areas are found in E. Morin, *Commune en France: la métamorphose de Plodémet* (Paris, 1967); J. M. Halpern, *Social and Cultural Change in a Serbian Village* (New Haven, 1956), and J. M. Halpern, *A Serbian Village* (New York, 1958).

mands. The range of agricultural goods required from domestic farmers varies greatly according to differing national policies for self-sufficiency, the conclusion of international trading agreements (drawn up for economic or more overtly political reasons), the

TABLE 6

Economically Non-viable Units as Percentage of All Farms

Austria	26	Italy	64
Denmark	15–30	Netherlands	50
France	53	Norway	58
Federal Germany	50	Portugal	50
Greece	53	Sweden	62–84
Ireland	50		

Source: O.E.E.C., *The Small Family Farm: A European Problem* (Paris, 1959), p. 21.

operation of techniques for financing agricultural protection (tariff barriers, subsidies, deficiency payments, etc.) and the changing food tastes of an increasingly urbanised European society. For these reasons agricultural production in Europe is

TABLE 7

Annual Rates of Decrease in Farm Numbers and in Agricultural Population in Western Europe, 1950–60

(percentages)

	Farms	*Agricultural population*		*Farms*	*Agricultural population*
Austria	0·8	3·5	Luxembourg	2·5	2·9
Belgium	2·6	4·0	Netherlands	1·4	2·4
Denmark	0·4	2·6	Norway	0·8	2·6
Federal Germany	1·6	4·2	Sweden	1·8	4·1
Iceland	0·7	2·8	United Kingdom	1·2	1·7
Ireland	0·9	2·2			

Source: Organisation for Economic Co-operation and Development, *Agriculture and Economic Growth* (Paris, 1965), p. 63.

still plagued by surpluses and shortages of different commodities which could be avoided if a fully modernised agricultural system operated.

Finally, there is the enormous question of deciding the rightful place of farming in the social, economic and land-use pattern of modern Europe. Technical progress, expressed in increasing farm yields, and continued freedom from international warfare have combined to release rural land from agricultural use. At the same time growing urbanisation, greater affluence and personal mobility are creating new demands for recreation space which will have to be met in the countryside.

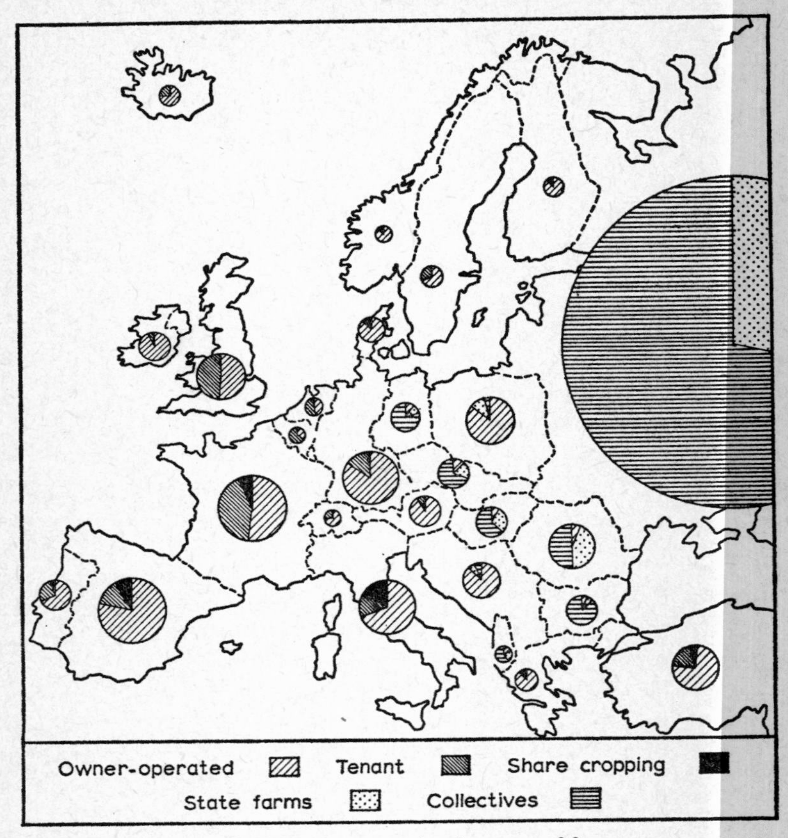

FIG. 8 Land occupation by type of farm

In order to reduce some of these problems and guide necessary changes, European governments have become increasingly concerned with the planning of agricultural activity since 1945. Initially the main aim was restarting production after the disruption of the war. This was then modified to increasing food supplies and ending rationing. Now many governments are trying to tackle problems of agricultural overproduction. But government intervention in farming is a highly explosive political issue involving the manipulation of food prices charged to increasingly numerous urban consumers and the provision of financial support for the contracting and often inefficient farming section of the community.

In Eastern Europe the establishment of Communist regimes involving the socialisation of the countryside permitted revolutionary government intervention in agriculture through land reform and collectivisation. Many common problems still exist in European agriculture as a whole, but the radical changes in land occupation in Eastern Europe have imposed fundamental contrasts between East and West (Fig. 8).

Collectivisation was introduced primarily for political reasons but also to satisfy social and economic objectives. It was hoped that the new agricultural enterprises would be welcomed by the farming population, would permit increased efficiency through radical changes in size and organisation of agricultural units, and would allow surplus rural population to be siphoned off into industry. Less thorough forms of intervention have operated in West European countries, sometimes with the clearly economic motive of streamlining agricultural production but more often with the ambiguous socio-economic aim of both improving agricultural efficiency and bettering the lot of the excessively large farm population. The establishment of the Common Market has required an attempt to harmonise national policies. The hard thinking which has resulted underlines basic faults in West European farming, which is grossly inefficient when judged against other sectors of the economy.

2. LAND REFORM AND COLLECTIVISATION OF EAST EUROPEAN AGRICULTURE

The establishment of Communist regimes in Eastern Europe permitted radical intervention which in theory might have allowed effective agricultural modernisation through amalgamating small family farms into massive collective and state farms. In fact success has been limited, largely through the unpopularity of these changes with some sections of the agricultural population, and in Poland and Yugoslavia the family farm has survived. For this reason it is essential to consider the recent evolution of farming in East European countries in some detail.

In spite of national differences, three common features were found in East European agriculture before 1939. First, there was high agricultural overpopulation, since lack of alternative jobs kept large numbers at work on the land. In the late 1930s agriculture supported between 50 per cent and 80 per cent of the population of each East European country save Czechoslovakia (38 per cent) and the eastern part of Germany (20 per cent). Second, farms of less than 5 ha predominated, ranging from 62 per cent of all units in Bulgaria to 85 per cent in Hungary. Third, there was a marked lack among the peasantry of the capital which would have been necessary if agricultural advancement were to be achieved.

Land reform had already been undertaken before 1939 and many large estates in Czechoslovakia, Yugoslavia, Romania and Poland had been confiscated between the wars and their land redistributed among the peasantry. Some large estates still remained. But with the exception of Hungary the peasants of Eastern Europe already owned the greater part of the land in 1945 and, except where governments acquired territory vacated by German farmers or the former ruling powers, the land available for redistribution was not extensive. Two important changes in land occupation have come about since 1945, however: the initial implementation of land reform which gave land to the tiller, and then collectivisation.

LAND REFORM

Land reform was carried through in every country between 1945 and 1948 and virtually all estates above limits ranging from 45 ha in Yugoslavia to 57 ha in Hungary were broken up. Compensation was available, save in East Germany, but only at a very low level. A few large properties remained intact to operate as state farms and research and teaching institutes. Normally the confiscated land was distributed among the peasantry and new farms of 5–10 ha were created. Such a change is illustrated by the situation in Hungary before and after land reform (Table 8). The new and slightly enlarged farms relied on family labour and had access to very small amounts of capital. They were not suitable for mechanisation or the application of scientific farming techniques. It had been politically desirable for the new regimes to distribute land among those who worked it, but the fragmentation of property which followed provoked problems of production and marketing and serious food shortages resulted.

Land reform in fact had been envisaged only as an intermediate step to win peasant support and sympathy for the new governments. The immediate introduction of Communist industrialisation policies would have provoked strong hostility from agrarian groups and the Church throughout Eastern Europe. Warriner has criticised the land reform as not being genuine since its object was neither to give the peasants permanent rights of ownership nor to help them become more effective producers.[5] The full implementation of socialist policies for agriculture required a

TABLE 8

Percentage of Farm Population in Hungary

	1935	*1946*
Landless labourers	32	11
Smallholders with less than 3 ha	50	61
Farmers with 3–12 ha	14	24
Farmers with more than 12 ha	4	4

Source: *World Atlas of Agriculture: Europe, U.S.S.R., Asia Minor* (Novara, 1969), p. 212.

second reorganisation. Farm labour needed to be moved from the land to new industrial jobs. It was also hoped to satisfy domestic food requirements and at the same time retain foreign markets. Achievement of these ends was considered by East European governments to be impossible with the system of peasant farming which land reform had strengthened.

As has already been shown, agricultural yields in post-war Eastern Europe had fallen disastrously from pre-war levels. Taking the example of wheat production, average yields in Poland had fallen from 1460 kg/ha (1937) to 950 kg/ha (1947) and in Hungary from 1490 kg/ha to 1020 kg/ha. Polish production was halved from 19·7 million quintals to 9·8 million quintals and that for Hungary toppled from 26·8 million quintals to 10·0 million quintals. For both political and economic reasons a radical reorganisation of East European agriculture was initiated in

[5] D. Warriner, *Land Reform in Principle and Practice* (Oxford, 1969), p. 14.

1947–8 when peasant farmers were enjoined to co-operate in forming larger enterprises. A few socialised elements had already been set up, such as machine and tractor stations in Bulgaria and Yugoslavia and model state farms in Romania and Hungary.

COLLECTIVISATION

Two main types of enterprise were established: state and collective farms. The former are owned by the East European states and financed by their centralised authorities. Their employees receive wages in the same way as industrial workers and profits are returned to national budgets. By contrast, collective farms are owned by their members, among whom profits are distributed partly in cash and partly in kind on the basis of the volume of work contributed throughout the year. Members of collective farms are allotted small patches of land, usually less than 0·5 ha, to cultivate for themselves and are generally permitted to keep a few livestock. Both state and collective farms are incorporated in the totally planned economies of Eastern Europe, guided by five-year plans. Their production is partly, in the case of collectives, or wholly, in the case of state farms, fixed by the planning authorities. Collectives are now the basic forms of land occupation in most East European countries. State farms cover much smaller areas, ranging from 6·4 per cent in Yugoslavia to 44·0 per cent in Romania (Table 9).

Collectivisation varied from country to country in details of

TABLE 9

Land Occupation in Eastern Europe by Farm Type

(percentage of total area)

	Owner-operated	*State farms*	*Collective farms*
Albania	20·0	7·5	72·5
Bulgaria	0·1	6·8	93·1
Czechoslovakia	11·1	21·1	67·8
East Germany	7·4	7·6	85·0
Hungary	3·7	32·6	63·7
Poland	86·0	12·9	1·1
Romania	4·8	44·0	51·2
Yugoslavia	87·6	6·4	6·0
U.S.S.R.	0·0	31·0	69·0

Source: *World Atlas of Agriculture: Europe, U.S.S.R., Asia Minor* (Novara, 1969), p. 8.

timing and approach between 1948 and 1952. The general reaction from the farming population was resistance, lack of co-operation and even sabotage. As Warriner puts it, 'farmers who own land are not willing to surrender it; even if they own very little, it is their only security in an insecure world'.[6] In addition to failing to obtain immediate popular support, collectivised farming lacked the necessary financial investment from state sources which such an enormous change demanded. Production failed to expand according to plan and demands for foodstuffs increased more rapidly than supplies. By 1952 an agrarian crisis affected virtually the whole of Eastern Europe. Collectivisation was slowed down or temporarily discontinued after Stalin's death in 1953, but was later resumed. There was no marked change in Bulgaria, however, where collectivisation had already gone further. Political changes in Eastern Europe led to further disruptions in 1956. These were temporary in Bulgaria and Czechoslovakia but were long-lived in Hungary and East Germany and virtually permanent in Poland.

The hostility provoked by collectivisation was summarised by Sanders in 1958: 'The crux of the problem now faced by the Communist regimes is fairly simple. The peasant, who for centuries has longed for land of his own or else has been attached to actual holdings passed down from his forefathers, has little relish for the regimented way of life represented in the collective farm. As a result, he works under protest and with little enthusiasm. Production is low and the regimes are forced to import foodstuffs into agricultural countries which exported these same commodities before the Second World War.'[7]

This resistance to collectivisation can be seen by the considerable amount of time and energy devoted by agricultural workers in Eastern Europe to their private plots, in spite of official criticism. Indeed, for the production of such commodities as vegetables, orchard fruit and livestock, the private plots are of predominant importance. The example of Hungary in 1961 provides an extreme illustration, but one which is comparable with the U.S.S.R. Private plots covered only 13 per cent of the arable area in Hungary but produced 33 per cent of wine, fruit and

[6] Ibid, p. 67.

[7] I. T. Sanders (ed.), *Collectivisation of Agriculture in Eastern Europe* (Lexington, Ky, 1958) p. 1.

TABLE 10

State Farms and Collectives in Hungary, 1949–61

	State farms		*Collectives*		*Workers*	*Proportion of arable*
	No.	*'000 ha*	*No.*	*'000 ha*	*in state and collective farms*	
1949	–	–	1367	182	36,000	2%
1952	501	920	5110	1500	369,000	45%
1956	466	967	2089	597	119,000	36%
1961	33	976	4566	7656	1,204,000	93%

Source: A. Blanc, P. George and H. Smotkine, *Les Républiques Socialistes d'Europe Centrale* (Paris, 1967), p. 185.

vegetables, 60 per cent of milk, 64 per cent of meat and 90 per cent of eggs. In the U.S.S.R. 45 per cent of meat and milk and 90 per cent of eggs came from private plots.

Faced with peasant opposition and unsatisfactory agricultural production, some East European countries realised that they could no longer afford collectivisation at the expense of reduced output. Thus to raise production they permitted a degree of de-collectivisation, allowing peasants to take over some of the land which formerly had been theirs. In October and November 1956 three thousand collectives were dissolved in Hungary, but since then their number has been increased and they now cover 93 per cent of the arable area (Table 10).

In addition to peasant hostility to collectivisation there were also problems stemming from governmental concern to enlarge agricultural units. Tiny family farms of 5 ha or less were certainly unsuitable for modern agricultural techniques, but the governments wanted bigness in agriculture without being prepared to make the necessary capital investments to give this bigness a reasonable chance of success. Agriculture became the Achilles' heel of all East European economies save that of Poland, where collectivisation had collapsed after 1952. In spite of a general relaxation in collectivisation for a time after the mid-1950s, all East European countries save Romania had to make contracts in the early 1960s to purchase large shipments of wheat from the United States, Canada and other Western countries.

NATIONAL DIFFERENCES IN COLLECTIVISATION

In spite of reversals, collectivisation of one form or another was completed everywhere in Eastern Europe by the spring of 1962 save in Yugoslavia and Poland. Now, a decade and a half after the political upheavals of the mid-1950s and the modifications in agricultural organisation which they provoked, three groups of countries may be distinguished on the basis of land-holding (Fig. 8): first, those where collectives dominate (Albania, Bulgaria and Romania); second, those with modified forms of collectives (Czechoslovakia, East Germany and Hungary); and third, those with important owner-occupation of family farms (Poland and Yugoslavia).

Collectivisation was accomplished rapidly in the first group. These backward countries lacked capital to modernise agriculture in the early 1950s and the short-term aim was clearly political. Thus collectivisation in Bulgaria, for example, was completed by 1957. A year later the collectives were regrouped into a thousand giant farms averaging more than 4000 ha each which replaced 1·1 million private farms. The pace of change had been slower in Romania, but the emphasis was turned abruptly towards collective farms in the early 1960s and by 1962 farming was completely collectivised. Romanian state farms average 3000 ha but some reach several tens of thousands of hectares. Following reorganisation of Romanian collectives in the 1960s only 2 per cent are now smaller than 500 ha compared with 40 per cent in 1961.

The social and political background of the second group of countries differed from the first. Collectivisation was discontinuous in Czechoslovakia, East Germany and Hungary after its dogmatic implementation in the early 1950s had been met with fierce resistance. Farmers in Czechoslovakia lost the incentive to increase production. They had been deprived of their land by collectivisation and were reduced to the level of labourers. Governmental manipulation of agricultural prices further cut back the rewards which farmers received. In 1953 a new policy was introduced, involving a reduction in the pressure to collectivise, an increase in income for farmers, and a greater allocation of productive resources to agriculture. But renewed government pressure to collectivise in 1956 meant that the previous benefits and goodwill were lost. General depression of the Czech economy between 1962 and 1964 provoked another reappraisal. Agricultural incomes were raised and individual collectives given greater freedom from central planning authorities to decide which crops should be produced and in what quantities.

East Germany long remained at the small-farm stage produced by the agrarian reform because until 1952 an overriding political objective was the unification of Germany. It was essential that nothing should be done in the East that would frighten family farmers in the West. Three types of collective farm were later established. The first involves only common tilling. In the second, collectivisation covers working the land and the use of animals, farm implements and other assets; livestock remains private property. The third type is based on the full collectivisation of all agricultural activities, property and livestock. Hungarian collectives have perhaps the most distinctive features in Eastern Europe. The most important is their economic independence. The farms have no obligatory plan or compulsory state delivery, being able to sell products as they wish. In 1963 the Hungarian Government pioneered the use of incentives in the collectivised system to give the farmers a share of their total production, regardless of whether planned targets were fulfilled or not.

In Poland and Yugoslavia, which make up the third group of countries, large-scale farming is represented almost entirely by state farms but these cover only a very small proportion of the agricultural surface since over 85 per cent is in private hands. Agrarian reform in both countries boosted the role of the family farmer. Polish reforms in 1944 led to the expropriation of estates of more than 50 ha (100 ha in west-central Poland) held by wealthy post-feudal landowners who represented only 0·2 per cent of the rural population. In all, 6·1 million ha were used to enlarge more than one million existing farms and create new holdings in western and northern Poland (Table 11). A maximum size of 15 ha was fixed but farms averaged only about 10 ha. In 1945 one-third of all Polish farms had been enlarged or created by land reform. In west-central Poland, and in areas from which German farmers had been evicted, estates and large farms were frequently kept intact as state farms, but these cover only 13 per cent of total farmland.

Collectivisation in Poland began in 1950 but the change of regime in 1956 halted and reversed it. The dual structure of the agrarian economy, with an extremely important family-farm component, was confirmed. In late 1964 there were 1246 collectives (Table 12) cultivating only 1 per cent of the farmland and accounting for 1 per cent of total output. The collectives which have survived are not organised as rigidly as in the U.S.S.R. or in the first group of East European countries. After 1956 Polish collectives were able to purchase their own tractors and equipment. State farms have also been reorganised when their special

TABLE 11

Groups Benefiting from 1944 Land Reform in Poland

(percentages)

	Proportion of beneficiaries	*Proportion of land distributed*
Former farm workers on landed estates	25·8	47·2
Other landless peasants	14·2	17·9
Small farmers: less than 2 ha[a]	21·3	10·9
2–5 ha	30·2	19·5
5–10 ha	5·8	3·6
Rural craftsmen	2·7	0·9

[a] Size of farm before land reform.

Source: J. Tepicht, 'Poland – 25 Years After: The Family Farm Dominates', *Ceres, F.A.O. Review*, II 6 (1969), 40.

ministry was abolished and they were placed directly under the Ministry of Agriculture.

A similar situation exists in Yugoslavia, where land reform in 1945 increased the number of small farms. Collectivisation was

TABLE 12

Collective Farms in Poland

	Number of farms	*Membership*	*Area* (ha)
1955	9076	205,200	1,867,000
1956	1534	31,600	260,000
1964	1246	29,200	219,000

Source: J. F. Karcz, *Soviet and East European Agriculture* (Berkeley and Los Angeles, 1967), p. 430.

enforced after the break with Cominform in 1948 but was achieved only after strong peasant resistance, expressed in diminished production, reduced livestock numbers and an increasing amount of land being left uncultivated. Forced collectivisation was discontinued in 1951 and a policy of easing off implemented. In 1953 there was a further step in the direction already taken in 1951, with membership of co-operative groups becoming voluntary. Farmers were allowed to leave at will and take their land with them. Massive withdrawals of land resulted and a policy of guaranteeing land for farmers who remained in the collectives had to be drawn up. A maximum limit of 10 ha for private holdings was fixed to achieve this end. In fact the average unit is only 4 ha in size. Nevertheless the ultimate goal of socialised farming remains in the Yugoslav constitution.

Such a fossilisation of family farming structures in both Poland and Yugoslavia slows down the chance for farm enlargement and agricultural modernisation. In the early 1960s new regulations were introduced to stop the further fragmentation of family farms in Poland. The right of inheritance was recognised only for those for whom agriculture was the main source of income. In 1967 agricultural authorities were given power under certain circumstances to take over farms larger than 5 ha and to grant compensatory pensions to elderly farmers. In 1968 a large consolidation programme was started. New legislation provides under certain conditions for obligatory sale of land at auction where the state has the right of compulsory purchase. These measures are similar to schemes already operating in parts of Western Europe. In spite of such action the situation in Poland and Yugoslavia contrasts with what has happened elsewhere in Eastern Europe where small strips have been replaced for the purpose of mechanised cultivation by large fields of 5–10 ha which are greater than the average family farm before 1945. Some enormous units have been produced, for example the 150–200 ha blocks of wheat surrounded by windbreaks on the Hungarian Plain. The socialist farms have been equipped with large barns and machine shops at selected rural centres. In Hungary, for example, many farmhouses have lost their old agricultural functions which have been transferred to the rural centres so that they remain simply as residences for farm workers.

The overall results of collectivisation appear not to have been as good as had been expected in the early 1950s. Lazarcik, working from the example of Czechoslovakia, demonstrated that 'in Western Europe and North America where private ownership has remained, agricultural output increased 52 per cent and 66 per cent respectively, between 1934–8 and 1963, while in Czechoslovakia output rose by only 6–8 per cent for the same period. . . . In other East European countries where agriculture was collectivised, the performance of agriculture was similar to that of Czechoslovakia. In Poland, however, where collectives account for only 1·1 per cent of agricultural land, the agricultural output rose 48 per cent between 1949 and 1961–2.'[8] Similarly in Yugoslavia total agricultural output in 1965–6 had risen by 63 per cent above the very low level of the early 1950s. Immediately comparable statistics are rarely available, but figures for eight main crops, meat and milk suggest that by the early 1960s the production levels for the mid-1930s had been regained in six East European countries but not in East Germany or Czechoslovakia (Table 13).

By the mid-1960s the problems of East European collectivised agriculture had become apparent. It was no longer adequate simply to raise agricultural production as had been the case in the 1950s. In most countries a stage had been reached where investment needed to be increased to develop capital-intensive forms of agriculture. This was urged in the national plans of Poland, Romania and Hungary, which envisaged the introduction into agriculture of industrialised production methods. Authorities everywhere now consider that the rate of growth of agricultural output and its economic efficiency depend to a decisive extent on the degree of technical equipment provided. They have admitted that during the course of rapid industrialisation in the 1950s and early 1960s they often failed to establish appropriate relationships between the farm sector on the one hand and on the other the main suppliers of industrial equipment and machinery to agriculture and the processing industry, trade and construction. These weak links in national economies constitute a severe handicap to the policy of intensifying and modernising agricultural production. Bottlenecks in the organisation of the agricultural system are apparent as the volume of production increases. New ministries and planning bodies are attempting to co-ordinate agricultural production with the task of satisfying food requirements. Attempts by the C.M.E.A. to integrate agricultural development in the Communist bloc to supply Soviet markets and to achieve economic rationalisation, for example by defining a 'primary producer' role for Romania, have been met by varying degrees

[8] G. Lazarcik, 'The Performance of Czechoslovak Agriculture since World War II', in J. F. Karcz (ed.), *Soviet and East European Agriculture* (Berkeley and Los Angeles, 1967) pp. 385–410.

TABLE 13

Changes in Commodity Production in Eastern Europe between 1934–8 and 1959–63

(1934–8 = 100)

	Alb.	*Bulg.*	*Czech.*	*E. Germ.*	*Hungary*	*Poland*	*Roman.*	*Yugo.*	*E. Eur.*
Wheat	219	127	108	83	82	136	147	150	151
Rye	233	28	60	89	44	111	59	99	94
Maize	124	169	230	18	144	118	138	118	135
Potatoes	1250	388	59	91	89	106	213	159	99
Apples	600	1256	110	148	776	117	58	196	168
Grapes	228	165	82	–	108	–	87	117	113
Sugar beet	1000	1183	143	99	302	161	724	431	173
Tobacco	500	277	54	54	100	533	278	243	219
Meat (beef, pork)	213	159	124	104	122	158	121	143	128
Poultry meat	150	168	78	200	110	167	139	125	128
Milk	113	107	78	109	128	118	195	127	114
All commodities	148	168	86	97	125	115	165	145	118

Source of data from which table has been prepared: G. Enyedi, 'The Changing Face of Agriculture in Eastern Europe', *Geographical Review*, LVII (1967), p. 368.

of resistance in the still strongly nationalistic East European countries.

As has already been shown, East European farming remains excessively in the hands of the elderly and female sectors of the labour force. In the 1960s a serious effort was made to improve the situation by redistributing national incomes more favourably towards the farming sector. The wages of collective farmers and state-farm workers have risen substantially and are rapidly approaching those of urban workers. This may induce young and well-trained people to enter or remain in agricultural employment.

The new regimes which entered power in Eastern Europe after the Second World War instituted changes of a magnitude which would be unthinkable in the West. But it is clear that, in the crucial initial stages of collectivisation at least, the social objective of winning support from the farming population was sacrificed in favour of installing a new political system and extending socialisation into agriculture. Reversals in collectivisation during the 1950s and the return to a family farming system in Poland and Yugoslavia may represent a satisfactory social solution for farmers in those countries, but such a situation acts as a brake on national economic advance which requires farm enlargement, mechanisation and the release of labour for employment in industrial and tertiary activities. Clear parallels may be drawn between the structural problems of these two countries and those of much of Western Europe. Agricultural modernisation in the West requires a radical reorganisation of farming, but action similar to that taken in Eastern Europe, although perhaps justifiable on economic grounds, would be unacceptable politically.

3. AGRICULTURAL CHANGE IN WESTERN EUROPE

Government intervention in West European farming has been spurred on by various motives. In the immediate post-war period the concern was to expand output. Now many governments are trying to cut it back. Other measures have aimed not only at the social objective of raising agricultural incomes, which lag behind those of other sectors of the economy, but have also attempted to keep food prices at reasonable levels for consumers and to make agricultural production more efficient.

Major constraints hindering agricultural rationalisation include the fact that farmers are too numerous and their properties too small and fragmented. Nevertheless many Western governments have operated policies to stabilise the *status quo* by attempting to raise the standard of living for all farmers and agricultural workers. In France the stated aim has been to achieve parity between agricultural incomes and those from other sectors. Other countries have not gone quite so far but have sought to guarantee fair and proper remuneration (Netherlands, Switzerland, United Kingdom) or to ensure that those in agricultural jobs participate fully in the general development of the economy (Austria, Federal Germany, Sweden).

IMPROVING FARM INCOMES : PRICE SUPPORT, GRANTS AND LOANS

Four types of measure have been introduced to try to raise farm incomes: first, those which directly support them; second, those which raise incomes through improving the efficiency of existing farms; third, those which alter the shape and size of farms and promote cession of uneconomic holdings; and fourth, those which aid lagging regions through renovating farming and other sectors of the economy. Intervention has been highly complicated, with individual countries pursuing their own policies. Only with the creation of the Common Market have attempts at policy harmonisation been made. These will be considered later.

Price supports form the principal means among the first group of measures directly affecting farm incomes. Import restrictions, the fixing of prices for certain commodities and numerous other interventions are included. In fact small farmers with low incomes generally consume the greater part of their produce and sell relatively little. Price supports may be of small significance in raising incomes substantially but may be just sufficient to keep farmers on the land. The net result will hinder solutions for improving outdated farming which can only result from reducing the agricultural population. Variants on support measures include differential prices and subsidies payable to operators of small farms or those located in upland areas. Generally speaking the results are of limited value. Social objectives are not fully achieved, since only slight rises in income are normally produced.

Neither are economic motives met, since the already excessive and inefficient farm population is induced to remain on the land.

The second form of intervention covers measures to raise the efficiency of existing farms through grants or loans for improvements, aiding the formation of co-operatives for purchasing and marketing, and operating advisory schemes. Whilst undoubtedly better than the first type of intervention, since some kind of increased efficiency is implicit, this type of action aids larger, better-informed farmers who know how to take advantage of such schemes. From a social point of view increased efficiency may enable farmers to enjoy a higher standard of living, provided of course that prices of agricultural commodities rise sufficiently. At the present time of increasing and unsaleable surpluses of farm products in Western Europe it is argued that even from an economic point of view it is no longer desirable to raise the efficiency of every farmer who applies for assistance.

STRUCTURAL CHANGE

The third type of intervention measure involves reorganising farm structures. If sufficiently radical this may offer a more satisfactory solution than the forms of intervention considered so far. Two major problems are covered by the term 'fragmentation', namely the division of land into small farms and the pulverisation of individual farms into tiny strips which may be scattered over a large area and are hence time-consuming and uneconomic to work. Mechanisation may be precluded or highly unsatisfactory. Fragmentation of holdings into tiny strips has originated in several ways. Some represents the fossilisation of open-field patterns inherited from historic systems of communal farming. Other causes include inheritance laws which demand an equal division of property between heirs, and piecemeal reclamation of farmland from the waste. In the mid-1950s at least half of Western Europe's farmland was in need of consolidation, ranging from 5 per cent in Denmark and Sweden to 50 per cent in Federal Germany and Spain and 60 per cent in Portugal. The advantages of consolidation are shown by the fact that output per hectare can be increased by 20 or 30 per cent.

Consolidation schemes should ensure the amalgamation of scattered plots into compact holdings around farmsteads, with internal divisions being kept to the minimum necessary for efficient management, the whole being large enough to provide an adequate living for the farmer and his family. Schemes exist in all West European countries, but they vary tremendously in approach, from very simple exchanges of parcels to ambitious programmes of rural management in which consolidation is included with the control of soil erosion, manuring, the building of new and improved roads, water management, and even industrial development, rural slum clearance and the construction of new houses.

The degree of initiative required for implementing consolidation schemes varies greatly. In France, Greece, Spain and Switzerland consolidation can only follow requests from the majority of farmers in the area involved, but in Federal Germany the Government may take the initiative in certain designated areas. Once property consolidation has taken place it is essential that further pulverisation should be avoided and hence legislation exists to this effect in many countries.

Federal Germany illustrates the problem. Land fragmentation is severe in the south-west and has resulted from the legacy of open fields and the equal division of land among heirs. Voluntary exchanges of strips between owners are encouraged and subsidised, but the area affected is small, being less than 5000 ha per annum. Property consolidation operates under plans for accelerated consolidation and intergrated structural reform. The importance of simple acceleration programmes has been stressed recently because of their rapidity and relatively low cost, but the classic integrated plan remains in more common use, with 244,000 ha being thus consolidated in 1967 by comparison with 40,000 ha by accelerated schemes. The task facing German agricultural planners is enormous. Some 5·89 million ha need consolidation for the first time and 2·80 million ha which have been consolidated already demand further reorganisation in the light of modern requirements. Given the annual average completion rate of 290,000 ha, it will take about thirty years to consolidate land to meet the requirements of farming in 1970. But today's requirements will be totally outdated by A.D. 2000 or any future date when further and far more radical reorganisation programmes will be needed! The slowness of structural change and the constraint which this represents for future farm modernisation characterised not only Federal Germany but the whole of Western Europe.

Table 14

Percentage of Farms under 10 ha in Size in West European Countries, 1960

Country	%	Country	%
Austria	64	Italy	89
Belgium	75	Luxembourg	40
Denmark	47	Netherlands	54
Finland	71	Norway	90
France	56	Portugal	95
Federal Germany	72	Spain	79
Greece	96	Sweden	61
Iceland	38	Switzerland	77
Ireland	49	Turkey	84

Source: Organisation for Economic Co-operation and Development, *Low Incomes in Agriculture* (Paris, 1964), p. 65.

Farm enlargement is another massive problem. In the late 1960s the average size of farm in the Common Market countries was only 11·7 ha. Table 14, covering non-Communist Europe in the early 1960s, shows that farms under 10 ha represented more than 70 per cent of all holdings in ten countries and more than 90 per cent in Greece, Norway and Portugal. It is clear that such farms are not capable of producing a reasonable income unless converted to an intensive activity such as factory farming or market-gardening. But it is not easy to determine what the desirable size of farm for the future should be. Many European states aim to create units for full-time working by one or two men without additional help other than that of their families. Absolute sizes are rarely quoted. In Warriner's words, 'the optimum size, i.e. the size which maximises output per man, is a variable dependent on several factors: the density of farm population expressed in the man/land ratio, the type of land use, determined on the one hand by the type of soil and on the other by the market; methods of production determined by the supply of capital and the level of technology. These conditions vary between countries, and some of them change in the process of development.'[9]

Special legislation in France, Federal Germany, the Netherlands, Norway and Sweden ensures that when farmland falls vacant it should be used to enlarge neighbouring holdings. But the results of natural change of this kind are slow. For example, it takes a decade to raise the average size of farms in the Common Market by a single hectare. Attempts to speed up enlargement

[9] Warriner, *Land Reform in Principle and Practice*, p. 38.

have been introduced in Austria, Federal Germany, France and the Netherlands where annuities and compensation are available to elderly farmers who retire from agriculture and allow their land to be used for enlarging neighbouring farms. Other schemes give financial assistance to younger farmers who elect to retrain for some other job and allow their farms to be used for restructuring operations. In France, for example, special annuities were paid to 123,000 elderly farmers in the five years after the scheme was started in mid-1963. Very often only slight increases in farm size resulted and the annuities have been criticised for failing to induce real economic advantages and for merely providing larger old-age pensions.

In addition to plot consolidation and farm enlargement, many areas require their settlement patterns to be remodelled so that farm buildings can be sited on their farmland rather than in overcrowded villages. Policies for this kind of change have long been implemented in Federal Germany, but resettlement is an expensive process. Besides constructing new farm buildings, public utilities must be provided, including hard-surfaced roads from the newly dispersed farmsteads to the village. It is less costly to provide utilities to new farms close to existing settlements than to those several kilometres away. But nearby locations are rarely ideal as a long-term solution to problems of inefficient production and land fragmentation. Resettlement in groups or in small hamlets is now preferred to completely isolated farmhouses. Such a policy reduces the cost of providing utilities and cuts down the serious social isolation experienced by families moved away from nucleated villages.

There has also been land reform in southern parts of non-Communist Europe, namely Italy and Greece. The objectives were of a social or political nature, to provide farms for landless agricultural workers and to enlarge existing smallholdings through the division of estates. The end result is often just the opposite of farm enlargement and will create serious problems for future generations of agricultural planners.

Inter-war governments in Italy had refused to accept that severe problems of agricultural poverty existed in the South. But in the late 1940s soldiers returning from the war were not going to remain passive and Communism threatened. Land reform was introduced after complex legislation in 1950 and provided for the acquisition and redistribution of land from estates above a threshold of 300 ha. The reform had the impossible aim of both modernising agriculture and providing adequate employment and decent living standards for the agricultural population. The authorities considered that land reform was not necessary in regions where agricultural modernisation was already under way.

In all 700,000 ha were redistributed, mainly in the South and the islands. Some 112,000 families, about half a million people, received portions of land between 1951 and 1962. Priority was given to the poorest members of the agricultural community. Some 44,500 casual labourers who had no experience as independent farmers received completely new holdings, with farmhouses located away from the densely packed nucleated villages (Table 15). These new farms varied from 4 ha of irrigated land to about 20 ha of poor hill land, with an average of about 11 ha. In addition 67,500 families received additional plots of 2·5–4·0 ha to supplement existing smallholdings. Expropriation had been generally restricted to poor, extensively cultivated land which had to be reclaimed, deep-ploughed and in some cases irrigated before the people could be settled in.

The relative success of the Italian land reform varied greatly between regions. In the Metaponto irrigated farms of about 5 ha have been highly successful, but great problems remain on new farms in Sicily and Calabria. With the wisdom of hindsight many aspects of the reform have been criticised. Costs of reorganisation have been high, averaging between $6000 and $9000 for each family installed. It has been argued that trying to create a landed democracy in southern Italy in the 1950s was unrealistic and that investment in new industrial jobs would have been more worth while. Other criticisms stem from the small size of many new

TABLE 15

Land Reform in Italy, 1962

	Number	*Area* (ha)	*Average size* (ha)
New farms created	44,533	474,459	11·0
Shares to enlarge existing farms	44,485	111,023	2·5
Parcels to enlarge existing farms[a]	23,046	93,090	4·0
Land given to institutions	—	3,024	–

[a] In Sicily.

Source: Warriner, *Land Reform in Principle and Practice*, p. 407.

farms. Units of 6–8 ha have proved quite inadequate to employ a family and provide sufficient income even in a good year. The physical environment (soil capability, climate, etc.) was not adequately studied before new farms were located and their size determined. Many new 5-ha farms in hilly terrain have been abandoned, but only after the expenditure of considerable sums on building farmhouses and providing utilities. By 1962, 15 per cent of assignees had abandoned their holdings, but now the proportion must be considerably higher. Many landless labourers who became small owner-operators after the reform were not intellectually prepared to manage their own farms. The new life proved difficult for many of them and a vast educational programme has been required. A final criticism stems from the fact that the creation of 5–10-ha farms through land reform has led to a serious pulverisation of property which runs directly contrary to what has been attempted by agricultural planners in other Western European countries. Vast new problems in southern Italy include establishing property-consolidation and farm-enlargement programmes and trying to get small farmers, proud of their newly found independence, to pool their resources in co-operatives.

REGIONAL MANAGEMENT

The final form of government intervention to raise incomes in farming is covered by the regional management schemes. It is no longer adequate to view agricultural problems in isolation. They must be considered in the context of regional, national and now increasingly international economies. Integrated plans for rural management have been implemented in Greece, Spain, southern France and southern Italy and the islands. The Spanish Badajóz Plan for the middle section of the Guadiana river provides a good example. Legislation in 1949 sanctioned the management of the valley linked to constructing the giant Cijara dam and five ancillary dams. Work was started in 1952 for completion in 1970. Some 130,000 ha have been irrigated and forty settlements established. Newly installed farmers have been trained in irrigation and modern farming techniques. The agricultural programme has been co-ordinated with schemes for afforestation and for the introduction of industrial employment.

If such schemes are to be effective, a serious appraisal of land potential has to be made so that appropriate solutions for future agricultural or alternative uses may be determined. In Sweden, for example, land is classified according to its most likely long-run use. Thus grants and loans for farm rationalisation are available only for land destined to remain in agricultural use. Poor land presently being farmed may be suitable for afforestation in the future and in such cases special grants are available. In the 1960s it has become increasingly apparent that measures of such a radical nature are necessary to replace the mass of protectionist policies which have shielded an inefficient, overpopulous and overproductive agricultural system in Western Europe since 1945.

4. FARMING IN THE COMMON MARKET

When the Common Market was set up on 1 January 1958 over 20 per cent (15 million) of its working population was engaged in farming but farmers contributed only about 8 per cent of the Community's Gross Product. The Rome Treaty (25 March 1957) had singled out agriculture for special consideration to produce a Common Agricultural Policy (C.A.P.) with the following, and unfortunately conflicting, aims. Agricultural productivity was to be increased through technical progress and rational development. A fair standard of living for the farming population was to be ensured. Markets would be stabilised and the consumer be guaranteed reasonable food prices.

The Community's farms contribute between 5·8 per cent (Federal Germany) and 14·4 per cent (Italy) of national incomes and employ between 6 per cent (Belgium) and 25 per cent (Italy) of the working populations of the various countries (Table 16). National conditions vary greatly, but foodstuffs and raw materials of animal and vegetable origin account for one-third of the Community's total imports from non-member countries, especially in the tropical world, and one-tenth of all its exports. One-sixth of the trade between member countries is in agricultural goods.

TOWARDS A COMMON AGRICULTURAL POLICY

As has already been mentioned, each member nation has operated its own price-support measures and policies to improve farm

TABLE 16

Agricultural Characteristics of the Common Market Countries, 1965

	Belg.	*Lux.*	*Neth.*	*Fed. Ger.*	*France*	*Italy*	*C. Mkt.*
Agricultural surface ('000 ha)	1671	135	2281	14,090	33,926	19,582	71,685
Percentage of workers in farming	6	14	8	11	18	25	17
Farmland (ha)/farm worker	7·8	7·1	5·4	4·6	9·4	3·9	5·8
Average farm size (ha)	8·2	16·0	10·2	10·0	17·8	9·0	11·7
Percentage contribution of farming to G.N.P.	6·6	7·1	9·3	5·8	8·9	14·4	8·6
Tractors/'000 ha farmland	36	53	39	79	30	19	37
Fertiliser used (kg)/ha of farmland	255	137	242	187	87	84	104
Number of agricultural advisers	710	13	2980	7443	4465	3222	18,883

Source: R. Froment and F. Gay, *Géographie Economique: l'Europe occidentale d'économie libérale* (Paris, 1970), p. 133.

structures and marketing since the Second World War. These policies varied considerably. Italy imported much of its food and did not therefore exercise the same degree of intervention in home production as France, which was more self-sufficient. Differing price and support levels ensured protection against imported foodstuffs in each country, but in such circumstances there was no way of getting farm produce to flow freely across frontiers. It was therefore essential to reduce national policies to a single Community policy for prices, protection levels and marketing.

Not surprisingly, the six governments made slow progress in putting Community policy above national interests. At a conference in Stresa in 1958 under the chairmanship of Dr Sicco Mansholt, the Dutch Vice-President of the E.E.C. Commission responsible for agriculture, clear differences of opinion were put forward. The French emphasised organised markets, the Germans stressed structural reform as a way of raising farm incomes, the Italians argued in favour of liberalising trade and abolishing subsidies, and so on. In spite of formidable differences of opinion, the following ambitious objectives were set down which reflected the conflicting aims embodied in the Treaty for agriculture: (i) to maintain a close correlation between policies for improving marketing and farm structures; (ii) to increase trade in farm products between members of the Six and with third countries, and to eliminate all quantitative restrictions; (iii) to reach a balance between supply and demand, avoiding surpluses and allowing specific agricultural regions to concentrate on crops which they produced best; (iv) to eliminate subsidies which distorted competition between individual member countries and regions; (v) to improve the rate of return on capital and labour; (vi) to preserve the family structure of farming; and (vii) to encourage rural industrialisation, to draw away surplus agricultural labour and eliminate marginal farms, and to give special aid to the poorer agricultural regions.

A three-stage progression towards an agricultural common market was envisaged. The initial exploratory stage would last three years. A second stage, when national policies and prices would be aligned, would be completed by 1 January 1970. The final stage would involve the complete integration of agricultural organisation in the Community. However, no indication of the mechanisms to be used to achieve these aims was given. Mansholt persuaded the governments of the Six to replace their individual import duties and restrictions by a single system of variable levies which was to be used to neutralise the effects of differences in price levels between the Community and the world market and also to bring prices in line between individual member countries. The levies were to be administered by the Commission and not from national capitals. At this stage the Common Market countries skirted round the politically explosive issue of common prices: these were only to be achieved by subsequent decisions over the transitional period. Producers in the Six were nevertheless protected from competition from lower prices on world markets, since import prices were raised by customs levies usually very close to the highest existing national price rather than the average. The alternative method of using deficiency payments was unacceptable because of its high cost and the difficulty of administering a system involving claims from six million farmers, many of whom were poorly educated. Other reasons favouring the variable levy included the fact that it was considered to provide less of a firm guarantee than would deficiency payments and hence might be used to move some of the surplus labour force out of agriculture. In addition the Common Market countries did not have traditions of cheap food and there was no incentive to provide it at cheap world prices by deficiency payments.

In spite of agreements in principle, little was achieved in getting the Six to agree to the detailed implementation of the levy system during 1961. But time was running out and the first phase was due to expire by 31 December 1961. A marathon meeting, stretching into January 1962, was required before agreement was finally reached by the Council of Ministers on proposals submitted by the Commission embodying the levy principle. These covered a common marketing policy for grains, pig meat, eggs, poultry meat, fruit, vegetables and wine. The C.A.P. could go ahead. There could be no return to separate policies, even though compromises and allowances would be necessary to get the Community system running.

Price support through the buying-in of certain products when prices fell below fixed levels was financed by the European Agricultural Guidance and Guarantee Fund (FEOGA) which, at this stage, was itself financed partly by national contributions and partly from the proceeds of external levies. Table 17,

TABLE 17

European Agricultural Guidance and Guarantee Fund: Balance to 31 December 1968

($ millions)

	Contributions		*Repayments*		*Beneficiaries and Losers*	
	Guar.	*Guid.*	*Guar.*	*Guid.*	*Guar.*	*Guid.*
Belgium	156	23	95	15	−61	−8
France	436	82	875	44	+439	−38
Federal Germany	538	87	168	56	−370	−31
Italy	413	64	306	150	−107	+86
Luxembourg	5	1	1	3	−4	+2
Netherlands	200	27	303	16	+103	−11
	1748	284	1748	284		

Source: S. de la Mahotière, *Towards One Europe* (Harmondsworth, 1970), p. 144.

covering the period up to 31 December 1968, shows how the contributions and receipts for market support were distributed among the Six, with France receiving by far the greatest support and Federal Germany being the largest contributor. A rational but highly expensive protectionist policy was implemented at the expense of the consumer to cushion the large, and hence politically important, peasant-farming population of the Six and at the same time to allow large and efficient farmers to reap a considerable profit.

These new policies eliminated distortions caused by national subsidies and import quotas and established common quality standards for the various commodities. The first two years of the C.A.P. saw the tentative beginnings of a gradual alignment of prices that had been proposed by the Commission. But progress was very slow, and in November 1963 Dr Mansholt proposed that the crucial issue of grain prices should be tackled in a single jump and harmonised for 1964–5. This met with considerable opposition from the Federal Republic, and it was only after threats by France to withdraw from the Common Market that the proposals were agreed in December 1964. These formed the key to all the other common price levels subsequently agreed, though the unified prices for grain did not come into force until 1 July 1967.

Emboldened by the success of earlier marathon meetings, the

Commission outlined proposals in the spring of 1965 for the future working of the Agricultural Fund, for the abolition of intra-Community duties on industrial goods (two and a half years earlier than planned), for directing proceeds from agricultural levies and the common external tariff on industrial goods directly into an automatically financed Community budget (which had formerly relied on annual national contributions), and for extending the supervisory powers of the European Parliament in line with the enormous financial responsibilities it would receive. This proposal proved unacceptable to the French who opposed any increase in the authority of the Commission. They then withdrew from the negotiations, and the whole of the Community came to a halt. In January 1966, however, a compromise agreement was reached, the Commission's proposals were discarded and the future arrangements for agricultural finance left in abeyance. Subsequently, however, agreement was reached to advance the date of the full customs union by eighteen months to 1 July 1968. In May 1966 a new scale of contributions to the guarantee section of the Fund was fixed for 1966–7. During the rest of the transition period a variable element (90 per cent of each country's levies on farm goods) was introduced to cover about one-half of the cost of the Fund. The remainder was to be met according to a fixed scale from national budgets.

In December 1969 a fourth marathon meeting took place. A complex formula for the future financing of all Community expenditure resulted whereby the Six will gradually pay into a common fund all customs duties on imports from non-member countries and part of the revenue from the value-added tax to be levied throughout the Community in 1972, as well as the levies on food imports that are already paid into the Farm Fund. From January 1975 the cost of the Common Agricultural Policy will be entirely met from the Community's own budget.

Until now the C.A.P. has concentrated on creating a common market for farm produce. Common prices are well above world prices. But little has been done by the Community to modernise or restructure West European farming. Expenditure from the guidance section of the FEOGA which in theory is devoted to structural improvements has in fact been used for land drainage, irrigation and the modernisation of marketing facilities. Such programmes are valuable in their own way but they are not what is required to create a smaller, and hence more realistic, number of farms. Every member state has major structural problems for which it is hard pressed to find funds. There is little willingness to finance programmes jointly through the FEOGA. But if current nationalistic attitudes were to change, the expenditure for guidance operations from the FEOGA could soar.

THE PROBLEM OF AGRICULTURAL SURPLUSES

Governments have found that supporting current agricultural price levels, storing the resulting stocks, and exporting the surpluses at subsidised prices has placed a great burden on the Community's Farm Fund and on national exchequers. The C.A.P. has taken on added responsibility and its operating costs rose eighty times between 1962–3 and 1968–9, now forming by far the most expensive single element of Community organisation. High guaranteed prices have encouraged even greater production of unsaleable goods. Money has been poured into the Farm Fund to prop up domestic prices, to destroy surplus fruit and vegetables, to store butter and subsidise agricultural exports. In 1969 the Community had to budget for some $2·3 billion to buy, store or dump unsaleable supplies as a means of internal market support. The Community's wheat surplus in 1969 alone was 4 million tons (12 per cent of normal production) and the accumulated surplus at the end of that harvest was 12 million tons.

However, the most immediate production problem facing the Common Market in the 1970s will be to deal with mounting surpluses of milk and milk products. Consumption is stagnant and is held back by high consumer prices. The stored surplus of butter (the so-called 'butter mountain') increased dramatically after 1967 to reach 400,000 tons in the autumn of 1969. Without lower prices, fewer cows or reduced world production the surplus threatens to reach gigantic proportions. This situation requires drastic action. Already a start has been made to reduce stocks. Measures involve heavily subsidised exports, the inclusion of dairy products in overseas aid programmes, and the development of outlets in the domestic animal-feed and other industries. It is hoped that such action will keep stocks down to 210,000 tons in 1971–2. Unlike wheat, whose consumption per head in the Community is falling, much more butter could be sold to domestic consumers if retail prices were reduced. Economically sensible proposals to cut guaranteed prices for milk, to subsidise the slaughter of dairy cows in small herds and to switch to beef production have been decried as anti-social and provocative and have been rejected.

A really savage reduction of commodity prices might cut production, but there is considerable evidence that the fiddling with prices which is politically tolerable does not. The obvious problem is the marginal farmer. If producer prices are cut he will not produce less but will work even harder to produce more and maintain his income. Surpluses will increase rather than fall. Another solution to the milk flood might be to switch dairy farmers into beef production, but this is not possible for small farmers who cannot finance the quite different time scales of fattening for slaughter. Such small farmers ought to be switched away from cattle rearing entirely.

This, however, is not going to be easily achieved. Current modifications in prices are attempting to move some small farmers from wheat into feed grains such as barley of which the Community is short. This may work, but past experience of manipulated prices in the mid-1960s was not convincing. The advent of common grain prices made wheat and barley production less profitable for German cereal farmers but more profitable for the French. Contrary to what had been expected, French cereal acreages barely rose at all but those in Federal Germany increased by 4 per cent between 1963 and 1969. It is clear that a simple cutting of farm prices would not solve overproduction problems. Sophisticated and repeated price cuts might do some good, but more radical and perhaps more ruthless measures are needed to combat overproduction. Farm labour forces are declining by 3 per cent per annum in Western Europe, but farm productivity per worker is increasing simultaneously by 7 per cent per annum. The cultivated area of the Common Market is 5 per cent less than it was ten years ago and the dairy herd is not significantly larger, but wheat and barley yields are 20–25 per cent higher, maize yields are up by 30–40 per cent, sugar beet by 20 per cent and yields from dairy cows by 15 per cent.

THE MANSHOLT MEMORANDUM

Over the last few years increasing attention has been given to putting forward ideas for a remedial restructuring policy. In

1968 Sicco Mansholt presented a hard-hitting report on farming in Western Europe and went on to suggest measures to rescue agriculture from being an antiquated, tradition-bound way of life and make it into a modern business. He showed that the farming population of the Six was being reduced by 500,000 each year during the 1950s and 1960s, but this was considered quite unsatisfactory since it has been the young people who have flocked away from the land, leaving the old and unambitious who were unlikely to modernise their farming activities. Systems of guaranteed prices were cushioning inefficient producers, thus keeping marginal farmers on the land and making West European farming even more uneconomic.

Mansholt's analysis showed that two-thirds of West European farms were less than 10 ha in size and three-quarters were too small to permit their labour force to be used rationally. Only 3 per cent (170,000) were over 50 ha. On average each farm worker in the Common Market managed 6 ha whilst his British counterpart managed three times that amount. Whichever aspects of West European farming are considered, the small, peasant scale of activity is over-represented. Four million farmers raise dairy cows in the Six, but two-thirds of them have herds of less than five cows. Only 75,000 farmers have more than twenty cows apiece. The general conclusion of the report was that at least 80 per cent of West European farms were marginal, given the technical and economic conditions of the late 1960s.

Mansholt therefore proposed three objectives for West European farming by 1980: to accelerate the drift from the land, to change farm sizes radically, and to balance out the supply and demand for farm products. It was argued that farming should be viewed simply as one among many economic activities rather than as a way of life. Mansholt envisaged that a total agricultural population of 5 million in the Six would be desirable in 1980. This would represent only one-quarter of the 1950 figure of 20 million which had since fallen to 15 million in 1960 and 10 million in 1970, some 12 per cent of the total labour force at that time. In the past the reduction of the agricultural labour force has not been planned and has resulted from natural decrease and the fact that more attractive wages were offered by industry. The task of reducing the agricultural labour force in the 1970s has become more difficult, since the hard core now remains. Unlike past changes which affected farm workers and members of farming families, almost half of the total number of *farmers*, many of whom actually own land, will have to disappear during the coming decade. Mansholt estimates that 2·5 million farmers and farm workers will have to be pensioned off and an equal number retrained for jobs in industry and services.

Mansholt argued that every effort should be made to divert the children of farming families away from agriculture to take up other jobs. A second form of action would involve encouraging the elderly to leave farming. In 1970 one million workers in agriculture were over the age of sixty-five and two million more will pass that age by 1980. Incentives in the form of annuities and pensions need to be intensified. Thirdly, efforts should be made to attract some of the younger people in farming to take on other jobs. Even if all these very difficult measures were to be put into operation, many optimists expect that by 1980 one-fifth of the remaining five million farm workers would still be on backward farms and an equal proportion on traditional though, hopefully, modernised family farms.

Following on this plea for a reduced labour force, Mansholt turned to the allied need to convert small peasant holdings into larger agricultural enterprises to practise modern techniques and operate according to development plans which would require official approval. Suggestions for the ideal size of enterprise were usually expressed according to the scale of activity. Thus, a suitable farm for 1980 would have 80–120 ha of cereals, or would raise 40–80 dairy cows, 150–200 head of beef cattle, 450–600 pigs, or 100,000 head of poultry each year. The magnitude of these targets is emphasised when one recalls that two-thirds of all farms in the Six were under 10 ha in size in the 1960s and two-thirds of all dairy farmers had less than five cows apiece.

The third line of attack was linked to supply and demand conditions. In the short term, Mansholt argued that the guaranteed prices of some grossly overproduced commodities such as sugar beet and milk should be slashed so that production be cut and, hopefully, some surpluses disposed of. In the medium term, marketing procedures would need reorganisation. In the long term the total cultivated surface of the Common Market would have to be reduced before 1980 by about 7 per cent, namely 5 million ha out of the total 71 million ha. This would involve removing from agricultural use an area greater than the total land surface of the Benelux countries. The Vedel report on French farming has recently suggested that far more than 5 million ha should be withdrawn from farming. The long-recognised underdeveloped parts of the Common Market (southern France, the Massif Central, Corsica, southern Italy and the islands) would probably bear the brunt of this reduction, whatever its magnitude. Mansholt suggested that one-fifth of the liberated agricultural surface should be used for creating national parks and other recreational areas. The remaining four-fifths should be devoted to afforestation. Financial assistance would help landowners who were willing to convert farmland to other uses provided that it was located in areas approved for alternative uses by Community planners.

Not surprisingly, the most general immediate reaction to Mansholt's proposals was one of hostility, especially from elderly European farmers and from associations defending the traditional family farm. Mansholt was called 'a second Stalin' by those who feared that thoroughgoing changes even comparable to collectivisation would be unleashed. To take just one example of complaint, the plan would prevent pensioned farmers from selling their land on the best market and would aid poor peasants less than existing national schemes since they would be required to allow their property to be used for recreation, afforestation or planned farm amalgamation. However, such shock treatment underlined the predicament of most West European farmers and promoted rational reflection and analysis, especially from young farmers' associations which responded sympathetically to the Mansholt Memorandum.

The six governments have all been dubious about the proposals. Several doubted that surpluses would actually be reduced. Others disliked the costs involved, probably between $4 and $5 billion a year for the period 1973–5, even though Mansholt's stated intention was to reduce total agricultural expenditure to only $2 billion per annum by 1980. The Germans have pointed out that France and Italy would have the utmost difficulty in employing those who would be encouraged to leave the land but would be too young for pensioning off.

Fifteen months after the proposals of December 1968 Mansholt announced a modified memorandum, couched in more delicate phraseology. General objectives remained the same, but six specific proposals were outlined for a five-year period. First, instead of the various size criteria for the farm of the future, it was suggested that a minimum gross output of \$10,000–\$20,000 per male farm worker each year should be retained. Individual farmers or groups wishing to work together and receive financial aid should submit development plans. Only on approval of such plans would financial aid be granted. Second, farmers planning to switch production from surplus products (wheat, dairy goods and sugar beet) to beef and coarse fodder grains should be given priority treatment, for example in the redistribution of land for farm enlargement. A slaughter premium would be paid if dairy farming were abandoned. Third, sums of at least the equivalent of \$1000 per annum should be available for those willing to abandon farming and release their land for whatever purpose suited local planning. Farmers under fifty-five years of age and releasing land would receive financial assistance amounting to at least eight times the rental value of their property. Owner-farmers leasing land for at least eighteen years for approved farm modernisation schemes would also receive subsidies. Fourth, no more land should be reclaimed for farming from the sea or from scrub or forest. Instead the Community would pay at least 80 per cent of afforestation costs. Land tax on property newly turned over to timber or recreation should be refunded for at least nine years. Fifth, advice and training should be made available more readily for those who chose to remain in the Community's modernised farming sector and also for those who decided to retrain for other jobs. Encouragement should be given for establishing co-operatives and producer groups to increase further the scale of farming activities.

The implications of the Mansholt proposals are enormous. Whilst providing measures for revitalising agriculture it is clear that rural poverty cannot be solved entirely by structural programmes within agriculture. Farming programmes need to be integrated into general economic, regional and social policies. The provision of alternative job opportunities for displaced farmers, agricultural workers and their families requires the maintenance of a high level of economic activity and employment, and the availability of suitable training and retraining facilities. Conclusions of this kind are being reached throughout Europe as most countries have entered or are about to enter the third phase of post-war agricultural development, namely that of overproduction. Sweden, for example, has implemented policies to cut back agricultural output for over a decade. This is a new problem whose solution is diametrically opposed to past objectives of increasing the agricultural area through land reclamation and generally raising total farm production.

Looking into the future, it is certain that urbanisation, or rather suburbanisation, will continue apace.[10] Increased affluence and mobility will combine with shorter working weeks to create new demands by city dwellers for recreational space in the countryside. At the same time the further application of scientific progress to farming will demand a comprehensive remodelling of farm structures and the introduction of new forms of land use to replace agriculture. Just how governments in Western and Eastern Europe alike will walk the tightrope between political expediency and social justice on the one hand and the hard facts of economics and technological advance on the other remains a matter of complete conjecture.

POSTSCRIPT

A fundamental change in the C.A.P. was agreed at a marathon meeting of agricultural ministers in March 1971. In addition to fixing prices for the coming season, the Six accepted joint responsibility for social and structural measures to reform farming. $1500 million will be available for this purpose in the next four years, quite apart from sums spent by individual countries. Restructuring programmes will be launched in countries where they are lacking. Normally 25 per cent of their costs will be met from the Agricultural Fund but in specially defined 'economically backward regions' 65 per cent will be covered. Farmers and agricultural workers between 55 and 65 will qualify for minimum annual pensions of $600 if they agree to leave the land. Younger workers will receive vocational training for jobs outside agriculture if they so wish. These and other basic measures will be applied throughout the Six but individual countries will be able to augment them to meet specific needs. All forms of intervention in agriculture will be harmonised in the future. This reorientation of the C.A.P. should cut back overproduction by marginal farmers.

[10] See the essay in this series by T. H. Elkins on *The Urban Explosion*.

FURTHER READING

Published material on recent trends in European agriculture is voluminous but extremely fragmentary, often dealing only with conditions in individual countries or regions. Only the most useful and accessible English-language sources are listed here. An excellent and detailed bibliography is contained in S. H. Franklin, *The European Peasantry* (London, 1969). Many advanced regional geography texts offer valuable introductions to local problems. Annual publications from the United Nations Organisation provide agricultural statistics but these are frequently of limited use for comparative purposes. The annual *Economic Surveys of Europe* (United Nations, Geneva) and *The State of Food and Agriculture* (Food and Agricultural Organisation, Washington, D.C.) offer very detailed discussions on agricultural topics for both Eastern and Western Europe. The various publications of the European Community Information Service, in particular the monthly journal entitled *European Community* and the *Newsletter on the Common Agricultural Policy*, contain much useful material on Common Market agricultural themes.

J. F. Brown, *The New Eastern Europe: The Krushchev Era and After* (London, 1966).

M. Butterwick and E. N. Rolfe, *Food, Farming and the Common Market* (Oxford, 1968).

G. R. Denton (ed.), *Economic Integration in Europe* (London, 1969).

J. F. Dewhurst, J. O. Coppock and P. L. Yates, *Europe's Needs and Resources* (London, 1961).

F. Dovring, *Land and Labor in Europe, 1900–50* (The Hague, 1956).

A. Duckham and G. Masefield, *Farming Systems of the World* (London, 1970).

R. Dumont and B. Rosier, *The Hungry Future* (London, 1969).

R. Dumont, *Types of Rural Economy: Studies in World Agriculture* (London, 1970).

G. Enyedi, 'The Changing Face of Agriculture in Eastern Europe', *Geographical Review*, LVII (1967), 358–72.

European Community Information Service, *The Common Agricultural Policy* (London, 1967).

G. W. Ford, 'The Common Market: Agricultural Structural Policies in the Member States', *Agriculture,* LXXIII (1966) 410–15

G. W. Hoffmann, 'Changes in the Agricultural Geography of Yugoslavia', in N. J. G. Pounds (ed.), *Geographical Essays on Eastern Europe* (The Hague, 1961).

International Labour Office, *Why Labour Leaves the Land* (Geneva, 1960).

J. F. Karcz (ed.), *Soviet and East European Agriculture* (Berkeley and Los Angeles, 1967).

A. Lambert, 'Farm Consolidation in Western Europe', *Geography,* vol 48 (1963), 31–47.

S. de la Mahotière, *Towards One Europe* (Harmondsworth, 1970).

A. Mayhew, 'Structural Reform and the Future of West German Agriculture', *Geographical Review,* LX (1970), 54–68.

Organisation for Economic Co-operation and Development, *Position of the Agricultural Hired Worker* (Paris, 1962).

——, *Problems of Manpower in Agriculture* (Paris, 1964).

——, *Low Incomes in Agriculture* (Paris, 1964).

——, *Agricultural Policies in 1966* (Paris, 1967).

——, *Agricultural Development in Southern Europe* (Paris, 1969).

I. T. Sanders (ed.), *Collectivisation of Agriculture in Eastern Europe* (Lexington, Ky, 1958).

J. Tepicht, 'Poland – 25 Years Later: The Family Farm Dominates', *Ceres F.A.O. Review,* II, 6 (1969), 39 42.

M. Tracy, *Agriculture in Western Europe: Crisis and Adaptation since 1880* (London, 1964).

United Nations, *Economic Council for Europe, 1947–62* (New York, 1964).

D. Warriner, *Land Reform in Principle and Practice* (Oxford, 1969).
World Agricultural Atlas: Europe, U.S.S.R., Asia Minor (Novara, 1969).
P. L. Yates, *Food, Land and Manpower in Western Europe* (London, 1960).

INDEX